2014 VexIQ Team 963A Reisenschein

A Tooth For A Tooth

A VexIQ Gear Handbook

Written by Rohit Narayanan, Angela Wei, Christopher Kang, Megha Narayanan & Matthew Cox

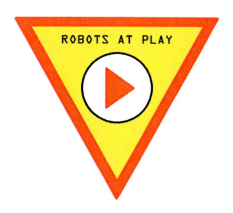

MRI Inspiration & Outreach Series

Copyright © 2014 by McLean Robotics Institute, McLean VA, USA

All rights reserved. Printed in the United States of America. Except for brief quotations embodied in reviews and critical articles, no part of this book may be reproduced in any manner whatsoever without the written permission of the McLean Robotics Institute, 963 Spencer Road, McLean, VA 22102. The McLean Robotics Institute is a nonprofit founded to advance and support excellence in STEM education.

Trademarked names may occasionally appear in this book. Rather than use a trademark symbol with every occurrence of a trademarked name, the authors use the names only in an editorial fashion and to the benefit of the trademark owner, with no intention of infringement of the trademark. **No affiliation, sponsorship, authorization or endorsement of this book or its contents by any organization or entity other than the McLean Robotics Institute is stated or implied.**

Library of Congress Cataloging-in-Publication Data

Narayanan, Rohit A. (2001-   )
    A Tooth For A Tooth: A VexIQ Gear Handbook / Rohit A. Narayanan (2001-   ), Angela M. Wei (2002-   ), Christopher H. Kang (2003-   ), Megha R. Narayanan (2005-   ) and Matthew J. Cox (2000-   )
Summary: A children's handbook of gearbox design showing compact and efficient gearboxes for robotics; includes a brief history of gears and simple explanations for younger children.
ISBN 1-499-11865-1 (paperback)
ISBN-13 978-1-499-11865-0 (paperback)
BISAC Categories: TEC009070 - Technology & Engineering / Mechanical
JNF051120 -Juvenile Nonfiction / Technology / How Things Work – Are Made
Library of Congress Control Number: 2014906958

Dedicated to

Tech-Savvy Kids - Engineers of the Future

TABLE OF CONTENTS

Acknowledgements vii
Preface - About This Book ix
1. An Introduction to Gears 1
    A. A Brief History of Gears 2
    B. What Is A Gear 5
    C. Types of Gears 6
    D. Rules of Gearing 7
    E. Gears In Your Life 8
2. A Practical Gearing Problem 11
    A. Our Gearing Problem 11
3. Mathematical Concepts Behind Gears 15
    A. Permutations and Combinations 15
    B. Simple and Compound Gears 16
    C. Stages of Gearing 16
    D. Torque and Speed 20
4. Gears Reductions and Gear Ratios 23
    A. Why Calculate Gear Reductions 23
    B. Gear Reductions and Gear Ratios 23
    C. Developing a Formula for Gear Reductions 25
    D. Worm Gears 29
5. Finding Compact Gear Permutations 31
    A. Minimum Spanning Beam 31
    B. Stack Height 33
    C. Finding the Most Compact Gearing 38
    D. Relationship Between Stack Height and MSB 39

Table of Contents

E. Practical Considerations	40
6. Complex Gear Systems	43
A. Nonlinear Stacks	43
7. Conclusions	47
Appendix A - References and Sources	49
Appendix B – Online Source Data	51
About The Team	52
About the McLean Robotics Institute	54

ACKNOWLEDGMENTS

This book is inspired by and in support of the inaugural Vex IQ Robotics Competition, Add It Up, organized by the Robotics Education and Competition Foundation.

The authors also wish to acknowledge the countless volunteers, coaches, parents and participants who make robotics competitions possible.

A special thanks to our loving and helpful parents who have given us the opportunity to learn about gears and write this book. We are also grateful to our parents for their helpful suggestions and assistance in the final proofreading of the book.

We are also thankful to Professor Malcolm Burrows and Greg Sutton of the University of Cambridge, UK, for their kind permission to use images of the *Issus Coleoptratus* in our book.

Preface - About This Book

Our team has built lots of robots this year, twelve at last count. As the process of revisions continues, we will probably build many more. And yet, not a single one will be perfect.

Early on, every time a robot would seem complete, we would then find that its arm would not lift the required weight, or it would go up and down at the wrong speed. After a lot of trial and error, we traced the problem to the gearboxes. We had used gears without fully understanding them, half thinking that they were used for increasing the height of the tower!

Later, even after we learned to build simple gear boxes, with gear reductions of 15 and 25, we found that the arrangement with the gear reduction 15 was nice and fast, but wasn't powerful enough to lift the required load.

Dear Younger Readers

I am Professor Gearhardt and I love to simplify!

Are all these *Equations* and *Formulas* getting ahead of you?

Just look for me, and I will explain everything.

We also found that the arrangement with the gear reduction 25 easily lifted the weight, but was a bit too slow for a game with a time deadline. We wanted to see if there was a gear reduction in the middle that would lift the required load and was nice and fast.

To become more systematic about our design, we decided to do more research into the design of gearboxes. We looked through 819 different gear permutations! Along the way, we learnt many things about gears and how to use them properly. The end outcome was that our robots are now much more efficient and effective.

This book is about what we learned. We hope this analysis helps other robotics teams in their search for the perfect gearbox.

CHAPTER 1
AN INTRODUCTION TO GEARS

Gears are a critical part of our everyday life. Without a gear and chain in your bicycle, you might still be riding a Penny Farthing, and it would take a lot more than a few scraped knees to ride that home. When you open your garage, gears multiply the torque of the lifting motor to raise the heavy door. At a drive-in, when you lower your car window to get a drink, gears inside the car door make the window go up and down by converting rotary motion into linear motion.

Can you imagine how much more difficult life would be without gears? Gears are hidden in everyday objects, even if you haven't paid much attention to them. You have gears in your watch, can opener, pencil sharpener, wind-up toys and much, much more! Now let's dive in into the mystery of gears.

Chapter 1

A. A Brief History of Gears

It is useful to understand where gears came from. Gears are one of the oldest inventions of mankind. Gears were used as far back as the 27th century BCE. The Science Museum of London has a model of a Chinese South Pointing Chariot. The human figure on the two-wheeled chariot maintains the position it was placed in no matter how much the chariot turns or moves. This device works not by using magnets, but by using gears. The figure changes its orientation based on the difference in rotation between the two wheels. If the chariot moves 30 degrees to the left, the figure moves 30 degrees to the right. This device is somewhat similar to a proportional line follower in a modern robot! This appears to be the oldest recorded use of gears.

Figure 1.1 Archimedean Screw

An Introduction to Gears

Gears have been used since ancient times. One of the older examples is the well-known Archimedean screw (Figure 1.1). The Archimedean screw is technically not a gear, as it does not mesh with another solid gear (see Section 1.B), though it is shaped like a worm gear. It uses a spiral ramp to lift water from rivers to farm lands. Its widespread use began in Hellenic Egypt in the 2nd century BCE and it was said to be invented by Archimedes. A minority view is that the "Archimedean" screw predated Archimedes and it was used in the Hanging Gardens of Babylon 350 years earlier (5th century BCE).

The Greeks used gears in waterwheels and clocks. In the fourth century BCE, the Greek philosopher Aristotle wrote that, "[the] direction of rotation is reversed when one gear wheel drives another gear wheel" (see Section 1.D). After the Greek Era, there appear to be no significant advances in the field of gears for a while. Possibly, the feudal barons were too busy forcing King John to sign the Magna Carta to work on gears. In the 15th-16th century CE, Leonardo da Vinci drew some sketches of gears from the Greek civilization in his sketch book.

> **History of Gears**
>
> You might think that gears are a recent invention, but they have been around for a long time. Almost 5,000 years ago, the Chinese used gears to make a little figure on a chariot always point in the starting direction, no matter how much the chariot moved or turned. It was just like a compass!

The next big development in gear technology came with the invention of profiled teeth in the 17th century. You might imagine that gear teeth are just sharp pointy triangles on a circle, but if gears were made like that, there would be a lot of clashing between teeth that would eventually cause them to wear out and break.

Also, the driven gear would not run at a uniform speed. In order to solve this problem called gear chatter, the profile of the gear teeth was changed into matched involute curves. An involute curve is a spiral whose radius increases by the circumference of the starting circle every full turn.

In the 19th century, the gear hobbing process was invented, significantly reducing the cost of gears. A hob is a cutting tool that can shape a rotating circular disk into a properly shaped involute gear.

Gears and machines fueled the Industrial Age and the growth in prosperity in the modern world. Today, gears are used in innumerable ways from kitchen and home appliances to industrial machines.

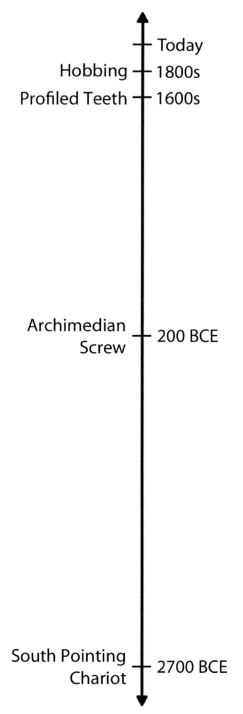

Fig. 1.2 Gear Timeline

An Introduction to Gears

B. What Is A Gear?

What is a gear? A gear is, simply, a wheel with teeth along its perimeter, although, as discussed above, it is quite a science to shape the teeth. Gears can be made out of almost anything. They come in many sizes and shapes including spur, bevel and worm (see Section 1.C). Gears can be categorized by shape, function, roles or size, but they are commonly categorized by the number of teeth they have or by their diameter.

Every gear plays one of three common roles. They are either driving gears, intermediate gears, or driven gears. Some intermediate gears are idlers. Idler gears don't change the gear ratio, but can be useful for changing the direction of rotation of a driven gear.

> **What is a Gear?**
>
> A gear is simply a disc with teeth around it. Not human teeth, but gear teeth. Gear teeth are sort of like spikes. Gears can come in many different shapes and sizes.

Fig. 1.3 Driving Gears, Driven Gears and Idlers

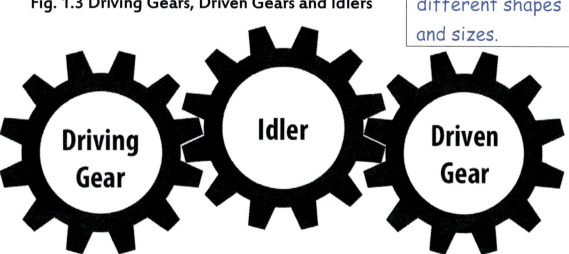

A VexIQ Gear Handbook

Gears can also change the speed and the torque of rotation. Torque is the power of the driven gear. When gearing increases the speed of rotation, it decreases the torque, and vice versa. For example, a gear train in a crane can help it lift a weight faster, or lift heavier weights while taking more time, depending on the gear ratio used. We will learn more about gear ratios in Chapter 4.

C. Types of Gears

There are many different types of gears. The most common types of gears are listed below:

- **Spur Gears** - A simple gear consisting of a cylinder with teeth projecting radially.
- **Worm Gears** - A gear that resembles a screw, that consists of a cylinder with one or more teeth wrapped around it spirally.
- **Bevel and Miter Gears** - Gears with conically shaped faces; gear shafts often do not rotate in a parallel manner.
- **Crown Gears** - Bevel gears with a 90 degree face angle.

Fig. 1.4 Worm Gear and Pinion

- **Helical Gears** - Spur gears with teeth not parallel to the axis of rotation.

- **Herringbone Gears** - Helical gears placed back to back to minimize side thrust.

An Introduction to Gears

- **Internal & Planetary Gears** - Gears with teeth on the inside rather than the outside
- **Sprockets & Chains** – Two sprockets that are not directly meshed are made to rotate by a chain. The chain allows the two sprockets to be far apart.
- **Racks** - A gear interface that produces linear motion from rotational motion. It is essentially just a toothed flat plate.

D. Rules of Gearing

Gears are orderly. They love to follow rules. Some common rules are listed below:

- When two gears mesh, and one gear rotates one tooth, the other gear will move one tooth, regardless of any differences in their diameters. We call this rule **A Tooth for a Tooth**. This fundamental law of gears is even older than Hammurabi's Law!
- In order for two gears to mesh, they must have the same tooth spacing. As discussed above, for them to move smoothly, the gear teeth have to have conjugate involute profiles.
- Gears can be connected in trains or in chains. A gear train is a series of meshed gears starting with a driving gear and ending with a driven gear. A gear chain is a series of meshed gears connected in a closed shape. A gear chain has one driving gear, and all the other gears in the connected shape, can be driven gears.
- A gear chain with an odd number of gears may

> **Rules of Gearing**
>
> There are many different types of gears, such as spur gears, worm gears, and bevel gears, and much more.
>
> Two or more gears with the same tooth spacing can mesh.
>
> Meshed gears can turn each other.

A VexIQ Gear Handbook

be able to mesh, but they cannot rotate.
- The ratio of output speed to input speed of a gear train (called the **gear ratio**), is the same as the ratio of the number of teeth on the initial driving gear to the number of teeth on the final driven gear.
- The ratio of output torque to input torque of a gear train (called the **gear reduction**), is the same as the ratio of the number of teeth on the final driven gear to the number of teeth on the initial driving gear.
- When gearing increases the speed of rotation, it decreases the torque, and *vice versa*. When the gear ratio is greater than one, the driven gear rotates faster than the driving gear. Thus, gear trains with gear ratios greater than one are used to increase the speed of rotation of a driven gear. When the gear reduction is greater than one, the driven gear rotates slower than the driving gear, while the driving torque increases. Thus, gear trains with gear reductions greater than one are used to increase the rotational torque of a driven gear.

E. Gears In Your Life

Now that you know where gears came from and what a gear is, let us consider examples in real life. In the bicycle example we considered earlier, pressing the pedal rotates a crank which in turn uses a chain to transmit power to the rear wheel. By increasing the gear ratio, we can make the bicycle move faster. We will get to gear ratios in Chapter 4.

A grandfather clock also uses gears. A series of gears connect to a drum, which connects to a cord which is weighted by a heavy weight. The force of gravity acting on the heavy weight turns the gears. The gears in turn, rotate the hands of the clock.

An Introduction to Gears

Gear shapes can also do other jobs. The pencil sharpeners that you can find on the walls of your classroom also have a special kind of gear called a planetary gear.

Did you know that gears are also found in nature? A type of plant-hopping insect, *Issus Coleoptratus*, has gears in its hips. The gears are only found in the young plant-hoppers and are used to propel them when they leap from plant to plant.

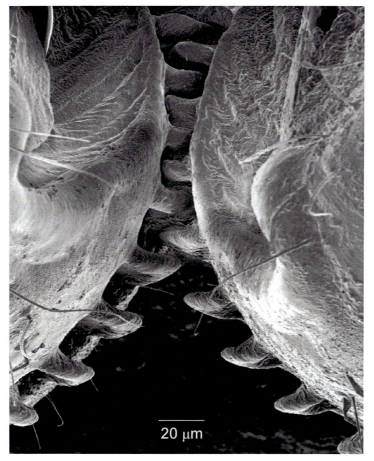

Fig. 1.5 Skeletal Gears of the *Issus Coleoptratus*

Gears In Nature

Gears are everywhere in your life - in clocks, bicycles, and more. Did you know that gears also exist in nature? The insect *Issus Coleoptratus* has gears in its legs and uses them to jump. When they grow up, the gears are lost. The Spiny Turtle's shell looks like a cogwheel gear.

Just before a young insect is ready to launch, the gears in its hips mesh with each other. With a swift motion, the *Issus Coleoptratus* jumps. The hip gears keep the legs in synchronization. The gears are lost when the nymphs grow into the adult phase.

The *Issus Coleoptratus* appears to be the only animal known to use gears. The *Heosemys Spinosa* (also known as the Spiny Turtle) has a shell that resembles a cogwheel gear, but it is only decorative. Some reptiles have cogwheel-like heart valves and some insects have gear-like knobs that make chirping sounds. However, none of these gear-like objects actually function as gears.

Fig. 1.6 Spiny Turtle

CHAPTER 2
A PRACTICAL GEARING PROBLEM

A. Our Gearing Problem

To construct a high performance robot, we need to find the correct balance between power and speed for our robot's arm. This is an interesting problem.

The key to its solution lies in gearing. However, this is not always obvious to a beginner. When gearing up a robot, they may not know enough to build an efficient and compact arm with the correct balance that fits the need. Worm gears, long arms, short arms - there seem to be no end of possibilities. To solve the problem more systematically, it is important to learn what gears really are, and how we could use them better.

We begin our study by exploring the gearing combinations that can be make using just:

    a. The three types of spur gears in the VexIQ set (the 12-tooth, the 36-tooth and the 60-tooth gears);
    b. No more than three stages; and
    c. Gearings that fit on a 4x12 plate, the largest plate in the VexIQ set.

Chapter 2

Fig. 2.1 Spur Gears In The VexIQ Set

> **The Problem**
>
> Now you know that are many types of gears and they are used for various purposes.
>
> We can also use gears to power and drive robots.
>
> The challenge is to figure out the ideal combination of gears that will lift a robot's arm with the right balance of power and speed.

The reason we are only considering spur gears is that we will be doing some calculations later on which will become very complicated with worm or crown gears. Also, there are no additional benefits to adding those types of gears: worm gears lose a lot of energy through friction, and crown gears offer little advantage other than changing the axis of rotation.

We also restrict our analysis to three stages (see Chapter 3 to learn what a stage of gearing is) because with four or more stages, we will likely lose a great amount of energy through friction.

We also limit our analysis to a single 4x12 plate so that all the axles of a gearbox fits on one VexIQ plate. This increases the rigidity of a gearbox. The 4x12 plate is the largest plate currently available in the VexIQ set.

A Practical Gearing Problem

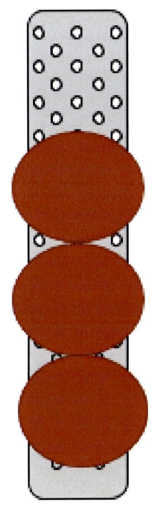

Fig. 2.2 Linear Gear Stack

In our research, idlers are not considered because they do not change the gear ratio, but can only make the tower taller or bulkier. Also, idlers make the Minimum Spanning Beam and Stack Height (see Chapter 5) harder to analyze and they also add unneeded friction.

We only analyze linear stacks (see Figure 2.2 at left), where all of the gears rotate with their axles aligned in a single plane. We do not consider non-linear stacks (see Figure 2.3 at right), because not too many of them exist and also as it would make it harder to calculate Minimum Spanning Beam and Stack Height. Non-linear stacks that work are also very hard to find with the VexIQ set. We will learn more about linear and non-linear stacks in Chapter 6.

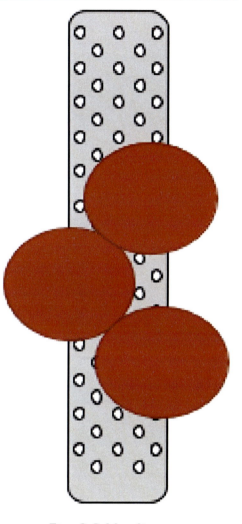

Fig. 2.3 Nonlinear Gear Stack

Chapter 3
Mathematical Concepts Behind Gears

Before we explain our research, here are some important mathematical concepts and definitions we will need to understand our analysis.

A. Permutations and Combinations

In English, we might say, "This salad is a combination of grapes, bananas and oranges." Most likely, it doesn't matter in which order the grapes, bananas and oranges are used. We might also say, "Here is the combination to the safe." In this case the order of the numbers is does matter, as a safe with the combination 123 would not work if you put in 312. In mathematics, we are more specific in differentiating between these two cases. If the order

> **Permutations and Combinations**
>
> A Permutation is a series of numbers whose order matters (like when opening a lock). For example, the sequences 1, 2, 3 and 3, 2, 1 are not the same.
>
> A Combination is a series where order does not matter (like the ingredients of a salad).

does not matter, we call it a **combination**. If the order does matter, we call it a **permutation**.

Here are (respective) formulae for permutations and combinations, where *n* is the number of things to choose from and we choose *r* of them, without repetitions.

$$P = \frac{n!}{r!(n-r)!} \tag{3.01}$$

$$C = \frac{n!}{(n-r)!} \tag{3.02}$$

B. Simple and Compound Gearing

When one input gear meshes with just one output gear, we call the arrangement a simple gearing. In compound gearing, the output gear of the first stage rotates a third gear that is on the same axis. This third gear meshes with a fourth gear (which is the also output gear of the second stage). This system of compound gearing can be extended to an infinite number of stages.

C. Stages of Gearing

The number of stages of gearing shows how many sets of simple gears are attached to the last free output axle. We use this concept to quickly find out how many permutations of gearing are possible.

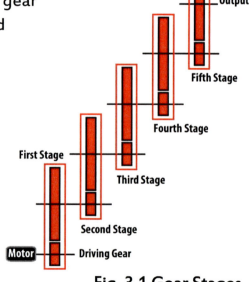

Fig. 3.1 Gear Stages

Mathematical Concepts Behind Gears

For simple gearing, where there is a driving gear and a driven gear, there would only be 9 permutations. However, when we consider compound gearing (gears on the same axle), the problem reaches an entire new level of complexity. You can see this in Figure 3.1 on the previous page.

If we restrict ourselves to three distinct sizes of gears, we can calculate number of permutations possible with every stage of gearing:

a. 1-stage gearing (Simple Gearing) has 9 permutations
b. 2-stage gearing has 81 permutations (9 first-stage x 9 second-stage)
c. 3-stage gearing has 729 permutations (9 first-stage x 9 second-stage x 9 third-stage)
d. 4-stage gearing has 6,561 permutations (9x9x9x9)
e. 10-stage gearing has 3,486,784,401 permutations (9x9x9x9x9x9x9x9x9x9)
f. In general, n-stage gearing has 9^n permutations

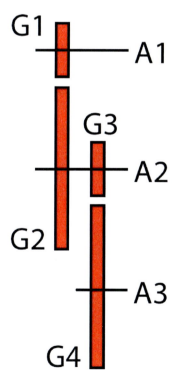

This is what it means to have different stages of gearing.

Assume that a gear G1, on axle A1, meshes with gear G2, on axle A2 as shown in Figure 3.2. Gears G1 and G2 have tooth counts T_{G1} and T_{G2}, respectively. Also assume that gear G1 rotates with angular velocity V_{G1}, at torque τ_{G1} while gear G2 rotates with angular velocity V_{G2} and torque τ_{G2}.

The relationships between the angular velocities and the torques of gears G1 and G2 are governed by the following equations:

Fig. 3.2 Compound Gear

$$V_{G2} = V_{G1}\left(\frac{T_{G1}}{T_{G2}}\right) \quad (3.03)$$

$$\tau_{G2} = \tau_{G1}\left(\frac{T_{G2}}{T_{G1}}\right) \quad (3.04)$$

This arrangement of gears is called a *one stage gearing* from G1 to G2.

Now assume that a gear G3, on axle A2, is rotating with gear G2, as shown in Figure 3.2. The angular velocity, V_{G3}, and the torque, τ_{G3}, of gear G3 are given by the following equations:

$$V_{G3} = V_{G2} \quad (3.05)$$

$$\tau_{G3} = \tau_{G2} \quad (3.06)$$

Let us now add a gear G4, on axle A3, to mesh with gear G3 as also shown in Figure 3.2. The angular velocity, V_{G4}, and the torque, τ_{G4}, of gear G4 are given by the following equations:

$$V_{G4} = V_{G3}\left(\frac{T_{G3}}{T_{G4}}\right) \quad (3.07)$$

$$\tau_{G4} = \tau_{G3}\left(\frac{T_{G4}}{T_{G3}}\right) \quad (3.08)$$

This arrangement of gears is called a *two-stage compound gearing* from G1 to G4.

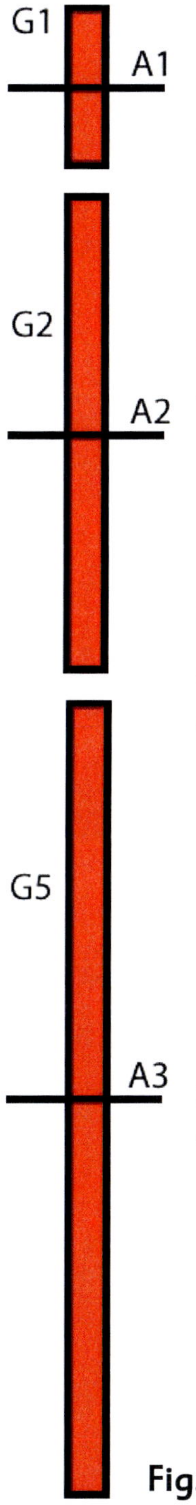

Mathematical Concepts Behind Gears

Substituting Equation (3.05) into Equation (3.07), gives us the angular velocity of gear G4 as a function of the angular velocity of gear G2:

$$V_{G4} = V_{G2}\left(\frac{T_{G3}}{T_{G4}}\right) \qquad (3.09)$$

Similarly, substituting Equation (3.06) into Equation (3.08), gives us the torque of gear G4 as a function of the torque of gear G2:

$$\tau_{G4} = \tau_{G2}\left(\frac{T_{G4}}{T_{G3}}\right) \qquad (3.10)$$

Now, substituting Equation (3.03) into Equation (3.09), we can get the angular velocity of gear G4 as a function of the angular velocity of gear G1:

$$V_{G4} = V_{G1}\left(\frac{T_{G1}}{T_{G2}}\right) \times \left(\frac{T_{G3}}{T_{G4}}\right)$$

$$= V_{G1}\left(\frac{T_{G1} \times T_{G3}}{T_{G2} \times T_{G4}}\right) \qquad (3.11)$$

Similarly, substituting Equation (3.04) into Equation (3.10), we can get the torque of gear G4 as a function of the torque of gear G1:

$$\tau_{G4} = \tau_{G1}\left(\frac{T_{G1}}{T_{G2}}\right) \times \left(\frac{T_{G3}}{T_{G4}}\right)$$

$$= \tau_{G1}\left(\frac{T_{G1} \times T_{G3}}{T_{G2} \times T_{G4}}\right) \qquad (3.12)$$

Fig. 3.3 Idler Gear

Now let us assume that gear G5, on axle A4, meshes with G2 as shown in Figure 3.3 on the left. G2 is an *idler* between G1 and G5. The angular velocity, V_{G5}, and the torque, τ_{G5}, of gear G5 are given by the following equations:

$$V_{G5} = V_{G2} \left(\frac{T_{G2}}{T_{G5}} \right) \tag{3.13}$$

$$\tau_{G5} = \tau_{G2} \left(\frac{T_{G2}}{T_{G5}} \right) \tag{3.14}$$

Substituting Equation (3.03) into Equation (3.13), gives us the angular velocity of gear G5 as a function of the angular velocity of gear G1:

$$V_{G5} = V_{G1} \left(\frac{T_{G1}}{T_{G2}} \right) \times \left(\frac{T_{G2}}{T_{G5}} \right) = V_{G1} \left(\frac{T_{G1}}{T_{G5}} \right) \tag{3.15}$$

Similarly, substituting Equation (3.04) into Equation (3.14), gives us the torque of gear G5 as a function of the torque of gear G1:

$$\tau_{G5} = \tau_{G1} \left(\frac{T_{G1}}{T_{G2}} \right) \times \left(\frac{T_{G2}}{T_{G5}} \right) = \tau_{G1} \left(\frac{T_{G1}}{T_{G5}} \right) \tag{3.16}$$

The presence of G2 didn't change the velocity or torque ratios between G1 and G5, so we can conclude that idlers have no effect on gear ratio or reduction. We therefore, do not need to consider idlers in further calculations.

D. Torque and Speed

We can combine and rewrite Equations (3.03) and (3.04) as:

$$\frac{V_{G2}}{V_{G1}} = \frac{\tau_{G1}}{\tau_{G2}} = \frac{T_{G1}}{T_{G2}} \tag{3.17}$$

Mathematical Concepts Behind Gears

If T_{G1}, the number of teeth on the input gear G1, is greater than T_{G2}, the number of teeth on the output gear G2, then τ_{G2}, the torque of the output gear G2, will be less than τ_{G1}, the torque of the input gear, G1. At the same time, then V_{G2}, the angular velocity of the output gear G2, will be more than V_{G1}, the angular velocity of the input gear, G1.

On the other hand, If T_{G1}, the number of teeth on the input gear G1, is lower than T_{G2}, the number of teeth on the output gear G2, then τ_{G2}, the torque of the output gear G2, will be more than τ_{G1}, the torque of the input gear, G1. At the same time, then V_{G2}, the angular velocity of the output gear G2, will be less than V_{G1}, the angular velocity of the input gear, G1.

Thus, when the torque decreases, the angular velocity, or speed, increases, and vice-versa, as illustrated in Figure 3.5 on the next page. The situation where the number of teeth on the output gear is greater than the number of teeth on the input gear leads to an increase in torque and a decrease in rotational speed. We call these gearings as having a gear reduction greater than one.

The opposite situation (where the number of teeth on the input gear is greater than the number

> **Torque and Speed**
>
> Imagine you are tightening a bolt. The rotating force needed to twist the bolt is called torque. How fast you twist is called the speed.
>
> You need the right combination of torque and speed to build the best robot arm. A robot arm with too little torque, cannot lift. Too much torque makes it too slow.

A VexIQ Gear Handbook

of teeth on the output gear) leads to an increase in rotational speed and a decrease in torque. We call these gearings as having gear reductions lesser than one.

We will discuss more about gear ratios and gear reductions in detail in the next chapter.

Fig. 3.4 Speed Vs. Torque

CHAPTER 4
GEAR REDUCTIONS AND GEAR RATIOS

A. Why Calculate Gear Reductions?

Now that we have all the concepts, let us go back to the specific question - how many gear reductions can be made with the three type of gears in the VexIQ set, with no more than three stages, that fit on a 4x12 plate? This is best done with a spreadsheet that looks at all the different permutations.

We first calculate the Gear Reductions of each of the permutations. It is an easy variable to analyze higher torques because they will always be higher than one, while Gear Ratios will always be decimal numbers between zero and one, and sometimes even in the thousandths.

B. Gear Reductions and Gear Ratios

Gear Reduction is the inverse of gear ratio. It is the relationship between the driven gear and the driving gear. Gear Reduction is the output gear's torque divided by the input gear's torque, and has an inverse relationship with the input

Chapter 4

to output tooth count. For example, if we have a 12 tooth driving gear and a 36 tooth driven gear, the gear reduction would be 3.

The Gear Ratio is the speed relationship between a driving gear and driven gear. The Gear Ratio is input speed divided by output speed, and has a direct relationship with the input to output tooth count.

If the driving gear has fewer teeth than the driven gear, we get increased torque and decreased speed. If the driving gear has more teeth than the driven gear, the opposite happens. So if we have a 12 tooth driving gear and a 36 driven gear, the gear ratio is 0.33 or 1/3.

Gear Ratios and Reductions

The Gear Ratio is the speed increase of a gearbox. The Gear Reduction is the torque increase.

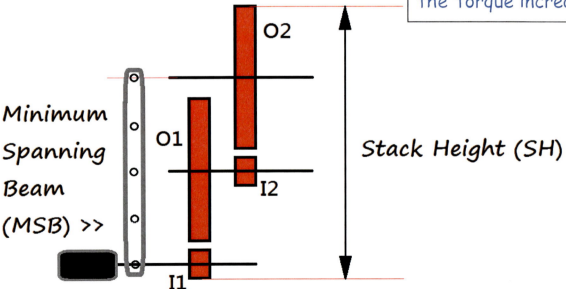

Fig. 4.1 Minimum Spanning Beam and Stack Height

Gear Reductions and Gear Ratios

C. Developing a Formula for Gear Reductions For Simple and Compound Gearings

If T_{O1} is the number of teeth on the first output gear and T_{I1} is the number of teeth on the first input gear, then the gear reduction, GR_{1S}, for simple gearing is simply the number of teeth on the output gear over the number of teeth on the input gear:

$$GR_{1S} = \frac{T_{O1}}{T_{I1}} \qquad (4.01)$$

The gear reduction formula for compound gearing is simply the product of all the individual gear reductions for each stage. So, the gear reduction, GR_{2C}, for two-stage compound gearing is:

$$GR_{2C} = \frac{T_{O1}}{T_{I1}} * \frac{T_{O2}}{T_{I2}} \qquad (4.02)$$

And as you may have guessed, the gear reduction, GR_{3C}, for three-stage compound gearing is:

$$GR_{3C} = \frac{T_{O1}}{T_{I1}} * \frac{T_{O2}}{T_{I2}} * \frac{T_{O3}}{T_{I3}} \qquad (4.03)$$

With a third stage, the gear reduction can range from below 0.01 to 125 with the three gears of the VexIQ set.

Though there are large number of possible gear reductions, as many as there are permutations, there are only a few unique reductions, as most of them are repeated many times. As illustrated in the graph below, for simple gearing, there are 7 unique reductions. For two-stage compound gearing there are 19 unique reductions. For three-stage compound gearing, there are 37. For each stage, there will be an equal number of reductions below and above 1. In statistical terms, 1 is the median.

Table 4.1: Number of Unique Combinations per Stage

Stages	Total Combinations	Unique Combinations
1	9	7
2	81	19
3	729	37

There seems to be a pattern here. However, further research will have to be done to find out what the pattern is. The authors are still looking into this.

Table 4.2 shows all the unique reductions from simple and compound gearing (up to 2-Stages). The ones in blue are the originals from the first stage. The ones in green are the new ones added in the second stage. The ones in red are the ones added in the third stage. They have been arranged to show that every gear reduction that is greater than one has a matching gear reduction that is less than one and the inverse of the other.

Table 4.2: Unique Gear Reductions for Simple and Compound Gearing

0.111	0.072	0.067	0.04	0.037	0.024	0.022	0.013	0.008	
0.12	0.185	0.2	0.216	0.333	0.36	0.556	0.6	0.926	
8.333	5.4	5	4.63	3	2.778	1.8	1.667	1.08	1
9	13.889	15	25	27	41.667	45	75	125	

Gear Reductions and Gear Ratios

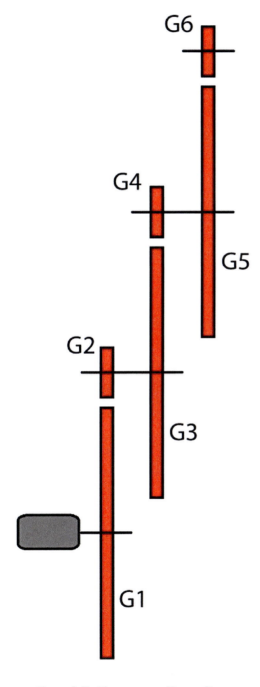

Fig. 4.2 Extreme Gear Ratio

Figure 4.2 shows the arrangement of gears that yields a gear ratio of 125, the highest gear ratio possible using the spur gears in the VexIQ set in a three-stage gearing. This setup will have very high speed. When the 60-tooth gear G1 turns once, the 12-tooth gear G2 turns 5 times. Gear G2 has the same angular velocity as gear G3 as they are both on the same axle.

Now, gear G3 has 60 teeth while gear G4 has 12 teeth, and so gear G4 will turn five times as fast as gear G3. Similarly, since gear G5 has 60 teeth and gear G6 has 12 teeth, they also share the 1:5 speed relationship.

Thus the three-stage compound gearing of gears G1-G6 will give a total three-stage speed increase of 125. We can also get this result by dividing the number of teeth in each input gear by the number of teeth in the associated output gear and then multiplying the results for each stage together to get 125, using Equation 4.03.

However, this gear arrangement may not work in practice (and it may even prove dangerous to try!). The maximum speed of a VexIQ motor is 120 rpm (revolutions per minute). If you multiply this by the gear ratio of 125 for this arrangement, you get an output

A VexIQ Gear Handbook

speed of 15,000 rpm. This is 1-1.5 times as fast as a large jet engine. The gears would simply shatter and fall apart if spun at those speeds. Also, if we had a gear ratio of 125, the associated gear reduction would be 0.008. This means that the gearing would not be able to bear any load at all. This gearing permutation likely would not be useful for any kind of real robot arm!

Of these combinations, some increase speed and some increase torque, though for any stage, there are always an equal number of reductions that increase torque and increase speed. For simple gearing, when the driven gear is larger than the driving gear it increases torque, while when the driven gear is smaller than the driving gear it increases speed. For compound gearing, it is not as easy to generalize, since the final value is the product of all the gear reductions. The driven and driving gear may have one ratio, but the intermediate gears will likely change that.

While gear combinations are important, other factors also impact torque and speed. Weight, length of the arm and friction can slow down a robot. Weight would slow down the robot because it would take more torque to move more weight. You would also need more torque to move a robot if it had a lot of friction. Shortening the arm would help increase speed.

Unique Gear Reductions

Although there are 729 ways to build a three-stage gearbox using 3 sizes of gears, only 37 of them are unique. 18 of these increase torque, while another 18 increase speed.

Gear Reductions and Gear Ratios

D. Worm Gears

Finally, let's look at gear reductions for worm gears. For the purpose of calculating gear reduction, you can think of the worm gear as a spur gear with one tooth, with the same measurements as the 12 tooth gear. However, there will be changes to the other variables we will be considering (see Chapter 5) as the worm gear is on a different plane.

In general, we do not use worm gears very much in making robotic arms, as friction loss takes away from any additional torque you gain. A worm gear drive sometimes takes up lots of space and does not rotate well with a 12 tooth gear.

Having calculated all of the gear reductions, it is surprising to find that there are a lot of repeated reductions. In the next chapter, we develop formulas to figure out the most compact permutation per reduction.

Chapter 5
Finding Compact Gear Permutations

Most times, when height is not a requirement, we want to find the most compact gear train for any gear reduction. To achieve this, two variables are important, Minimum Spanning Beam and Stack Height.

A. Minimum Spanning Beam

The Minimum Spanning Beam (MSB) is the shortest beam that can span all the axles of a gearing permutation. It is measured in units of holes. In the VexIQ kit, holes in plates and beams are spaced 0.5 inches apart. The diameters of the three sizes of spur gears in the VexIQ set are shown in Table 5.1 below (in inches).

Large (60 Tooth)	2.5 inches
Medium (36 Tooth)	1.5 inches
Small (12 Tooth)	0.5 inches

Table 5.1: Diameter of Spur Gears in VexIQ Set

Chapter 5

Let R_{I1} and R_{O1} be the radii (measured in hole spaces) of the first stage input and output gears, respectively. Let the input gear be on an axle A1 and the output gear be on an axle A2. The distance between the two axles is the sum of the radii of the two gears, I1 and O1. A beam that spans both these axles will need to have one more hole than the number of hole spaces between the axles. The MSB of a simple one-stage gearing (measured in holes), is given by the equation below:

$$MSB_{1S} = R_{I1} + R_{O1} + 1 \tag{5.01}$$

Further, assume that a second-stage input gear I2 is also on axle A2 and while the second-stage output gear O2 is on a third axle A3. Let R_{I2} and R_{O2} be the radii (measured in hole spaces) of the second-stage input and output gears, respectively. We already know the distance between axle A1 and A2. We now have to calculate the distance between axles A2 and A3. We can do this in a similar fashion to the first stage calculation above, by adding the radii of the second-stage gears I2 and O2. Thus, the MSB of a two-stage compound gearing, measured in holes, is given by the following equation:

$$MSB_{2C} = (R_{I1} + R_{O1}) + (R_{I2} + R_{O2}) + 1 \tag{5.02}$$

By extension, the equation for MSB for three-stage compound gearing is:

$$MSB_{3C} = (R_{I1} + R_{O1}) + (R_{I2} + R_{O2}) + (R_{I3} + R_{O3}) + 1 \tag{5.03}$$

where R_{I3} and R_{O3} are the radii of the third stage input and output gears, measured in hole spaces, respectively. We simply add up the radii of each gear pair of each gear stage and then add one to change the unit of measurement from hole spaces to holes.

Finding Compact Gear Permutations

In the VexIQ set, the radius of a gear in hole spaces is simply its number of teeth divided by 24. For example, if you divide the 60 teeth of a large 60-tooth gear by 24, you get 2.5. We can prove that the radius of a 60-tooth gear in hole spaces is 5 because its radius in inches is 1.25 and a hole space is 0.5 inches.

To make it easier to calculate when we do not have the exact radius, the MSB equations can be rewritten using only the number of teeth of the various gears. This is how it was written in the spreadsheet. Let T_{I1} be the number of teeth of the first stage input gear, and T_{O1} be the number of teeth of a first stage output gear.

$$MSB_{1S} = \frac{(T_{I1}+T_{O1})}{24} + 1 \qquad (5.04)$$

The equation for MSB for two-stage compound gearing is (in usual notation):

$$MSB_{2C} = \frac{(T_{I1}+T_{O1}+T_{I2}+T_{O2})}{24} + 1 \qquad (5.05)$$

Finally, the MSB for three-stage compound gearing is shown below:

$$MSB_{3C} = \frac{(T_{I1}+T_{O1}+T_{I2}+T_{O2}+T_{I3}+T_{O3})}{24} + 1 \qquad (5.06)$$

B. Stack Height

The Stack Height is the distance from the bottom of the lowest gear to the top of the highest gear. In theory, the calculation of Stack Height should be simple as in normal circumstances, the final output gear would stick out the highest and the first input gear would stick out the lowest in a gear stack. However, this observation works only for simple gearing.

The way we calculate the Stack Height for simple gearing is by adding the diameters of gear I1 and O1. The Stack Height of a simple gearing, measured in hole spaces, is given by the equation below:

$$SH_{1S} = 2R_{I1} + 2R_{O1} \tag{5.07}$$

where R_{I1} is the radius of the input gear I1, etc.

As we did for the Minimum Spanning Beam, the formula for Stack Height for simple gearing can be written in terms of the number of teeth. If T_{I1} is the number of teeth in input gear I1, and T_{O1} is the number of teeth of the associated output gear O1, the Stack Height of a simple gearing is given by the following equation:

$$SH_{1S} = \frac{(T_{I1} + T_{O1})}{12} \tag{5.08}$$

Table 5.2 below lists the Stack Height in hole spaces (HS) of each of the 9 permutations of simple gearings.

Table 5.2: Stack Height of Simple Gear Permutations

Size of Driving Gear In Teeth	Size of Driven Gear In Teeth	Stack Height in Hole Spaces (HS)
60 tooth gear	60 tooth gear	10 HS
60 tooth gear	36 tooth gear	8 HS
60 tooth gear	12 tooth gear	6 HS
36 tooth gear	60 tooth gear	8 HS
36 tooth gear	36 tooth gear	6 HS
36 tooth gear	12 tooth gear	4 HS
12 tooth gear	60 tooth gear	6 HS
12 tooth gear	36 tooth gear	4 HS
12 tooth gear	12 tooth gear	2 HS

Finding Compact Gear Permutations

However, as we get to two- and three-stage gearings, the top gears of earlier stages are sometimes taller than the final output gear. As shown in Figure 5.1 below, in some two-stage cases, the first-stage output gear O1 may sometimes top the second-stage output gear O2. Also, as shown in Figure 5.2 below, in some two-stage cases, the bottom of the second-stage input gear I2 may sometimes be lower than the first-stage input gear I1. Luckily, for us, it is pretty rare to find such a permutation in the VexIQ set. However, the calculations below are not limited to the VexIQ gear set, but may generally be used for any gear set.

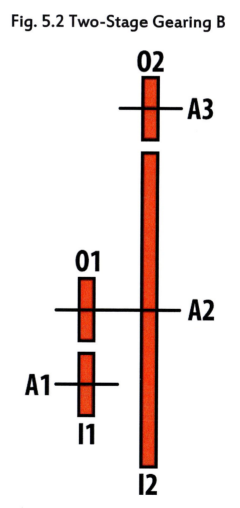

Fig. 5.1 Two-Stage Gearing A   Fig. 5.2 Two-Stage Gearing B

To calculate the Stack Height of a two-stage gearing, we have to sum the maximum distance above axle A3 with the maximum distance below axle A1 and then add both to the MSB for that arrangement (calculated in hole spaces). The maximum distance above axle A3 is the greater of (i) the radius of gear O2, or (ii) the radius of gear O1 minus the distance between the axles A2 and A3.

The maximum distance below axle A1 will be the greater of (i) the radius of first-stage input gear I1, or (ii) the radius of the second-stage input gear I2 minus the distance between the two axles A1 and A2. The distance between the axles A1 and A2 in hole spaces is just one less than the MSB for the simple I1-O1 gearing.

The formula for the Stack Height for two-stage compound gearing is:

$$SH_{2C} = MSB - 1 + MAX(R_{O1} - (R_{O2} + R_{I2}), R_{O2}) + MAX(R_{I1}, R_{I2} - (R_{I1} + R_{O1})) \quad (5.09)$$

To calculate the Stack Height of a three-stage gearing, we have to sum the maximum distance above axle A4 to the maximum distance below axle A1 and then add both to the MSB for that arrangement (calculated in hole spaces). The maximum distance above axle A4 is the greatest of (i) the radius of gear O3, or (ii) the radius of gear O2 minus the distance between the axles A3 and A4, or (iii) the radius of gear O1 minus the distance between axle A2 and A4.

The maximum distance below axle A1 will be the greater of (i) the radius of first-stage input gear I1, or (ii) the radius of the second-stage input gear I2 minus the distance between the two axles A1 and A2, or (iii) the radius of gear I3 minus the distance between axle A3 and A1. The distance between the axles A1 and A4

Finding Compact Gear Permutations

in hole spaces is just one less than the MSB for that three-stage compound gearing.

The formula for Stack Height for three-stage compound gearing is given by:

$$SH_{3C} = MSB - 1 +$$
$$MAX\ (R_{O1} - (R_{I2} + R_{O2} + R_{I3} + R_{O3}),$$
$$R_{O2} - (R_{I3} + R_{O3}), R_{O3}) +$$
$$MAX\ (R_{I1}, R_{I2} - (R_{I1} + R_{O1}),$$
$$R_{I3} - (R_{I1} + R_{O1} + R_{I2} + R_{O2}))\ \textbf{(5.10)}$$

The equation for two-stage Stack Height (expressed in terms of the number of teeth, like for Minimum Spanning Beam), is as follows:

$$SH_{2C} = MSB - 1 + MAX\left(\frac{T_{O1}-(T_{I2}+T_{O2})}{24}, \frac{T_{O2}}{24}\right) +$$
$$MAX\left(\frac{T_{I1}}{24}, \frac{T_{I2}-(T_{I1}+T_{O1})}{24}\right) \quad \textbf{(5.11)}$$

The formula for the Stack Height for three-stage compound gearing in term of the number of teeth is:

$$SH_{3C} = MSB - 1 +$$
$$MAX\left(\frac{T_{I1}}{24}, \frac{T_{I2}-(T_{I1}+T_{O1})}{24}, \frac{T_{I3} - (T_{I1}+T_{O1}+T_{I2}+T_{O2})}{24}\right) +$$
$$MAX\left(\frac{T_{O1} - (T_{I2}+T_{O2}+T_{I3}+T_{O3})}{24}, \frac{T_{O2}-(T_{I3}+T_{O3})}{24}, \frac{T_{O3}}{24}\right) \quad \textbf{(5.12)}$$

C. Finding the Most Compact Permutation for Each Gear Reduction

Generally, when we are trying to find the most compact permutation, we look at the MSB as we care about how long the plate will have to be rather than how tall the gear stack would be. However, if we were trying to fit a robot to a height limit (as is typically done robotics competitions like the VexIQ Challenge), it would not help to have a gear sticking up 2.5 inches in the air even if our MSB is below the height limit.

Tables 5.3 & 5.4: Range of Possible MSBs for each Gear Reduction

MSB	Gear Reduction			
	1	1.667	3	5
	1	0.6	0.333	0.2
2	1			
3	1		1	
4	2		2	1
5	4	1	3	2
6	10	4	2	3
7	5	9	9	4
8	18	2	4	18
9	4	18	21	2
10	37	2		18
11	1	21	15	
12	18			18
13		9	9	
14	9			3
15		3		
16	1			

MSB	Gear Reduction					
	1.8	2.778	8.333	9	15	25
	0.556	0.36	0.12	0.111	0.067	0.04
2						
3						
4						
5				1		
6				3	2	
7	2				6	1
8	9		2	3		3
9		1	9		9	
10	5	9		9		9
11			9		9	
12	9	3				3
13			6			
14		3				
15						
16						

Finding Compact Gear Permutations

Depending on the problem we are trying to solve, we can order the permutations in each gear reductions by one of these variables. The range of MSBs for each of the Gear Reductions is shown in Tables 5.3-5.5 for one-stage, two-stage and three-stage gearings. Gear Reductions are shown in green (first stage), blue (second stage), and red (third stage).

Table 5.5 Possible MSBs for All New Third Stage Gear Reductions

MSB	Gear Reduction								
	1.08 / 0.926	4.63 / 0.216	5.4 / 0.185	13.889 / 0.072	27 / 0.037	41.667 / 0.024	45 / 0.022	75 / 0.013	125 / 0.008
2									
3									
4									
5									
6									
7					1				
8							3		
9			3					3	
10									1
11	3					3			
12				2					
13		1							
14									
15									
16									

D. Relationship between Stack Height and MSB

As an additional exercise, we can graph the relationship between MSB and Stack Height. Through this graph, we can prove that Stack Height can never be smaller

A VexIQ Gear Handbook

than MSB (measured in hole spaces). This is because MSB is measured from axle to axle and every gear always sticks up beyond the axle it is on. Using the data below, we also find the highest and lowest Stack Height for every Minimum Spanning Beam. For example, with a MSB of 5 Holes, the lowest Stack Height would be 3 HS and the highest would be 8 HS.

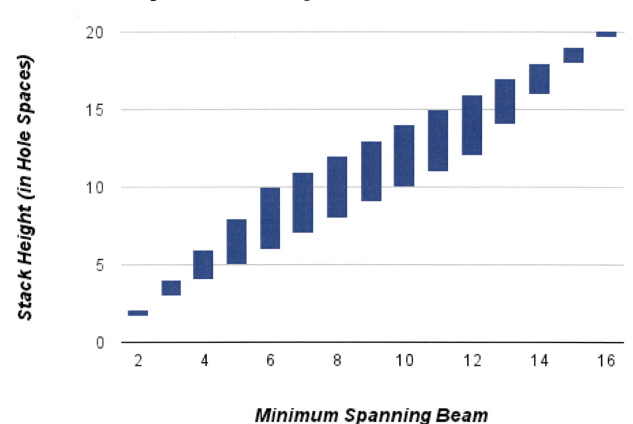

Fig. 5.3 How Stack Height Varies With MSB

E. Practical Considerations

We need to keep in mind that we are trying to find those gearing permutations that fit on the biggest plate in the VexIQ set (which 12 holes tall by 4 holes wide).

Finding Compact Gear Permutations

On the same plate, there are also offset holes in a 3 by 11 pattern. It is common to put axles in these half holes, and so, 11 holes would be the highest MSB you would practically use. If the MSB were 12 or more, we would not be able to securely attach a VexIQ motor on the same plate.

> The Minimum Spanning Beam (MSB) is the shortest beam that can cover all of the axles of a gearbox. For a simple gearing, the MSB is the distance between the two axles measured in Holes.
>
> The Stack Height is the total height of a gearbox. For a simple gearing, the Stack Height is the sum of the diameters of the two gears. It is measured in Hole Spaces.
>
> One can convert Hole Spaces into Holes by just adding one to Hole Spaces.
>
> The formulas for MSB and Stack Height are easy to understand and simple to calculate with a spreadsheet-type program. Once you calculate the MSB you can then calculate the Stack Height.

A VexIQ Gear Handbook

Chapter 6
Complex Gear Systems

A. Nonlinear Stacks

So far, we have only looked at linearly stacked gears. However, if we go beyond linear stacks, we might get more compact gearboxes. It is not easy to figure out the holes in which to place the axles of a non-linear gear stack. One of the ideas used initially is to attach a beam to one corner of a 4x12 plate and then rotate it slowly. The attachment corner is marked in red in Figure 6.1 below. We then look for holes on the plate that align perfectly with holes in the beam at various angular orientations of the rotating beam.

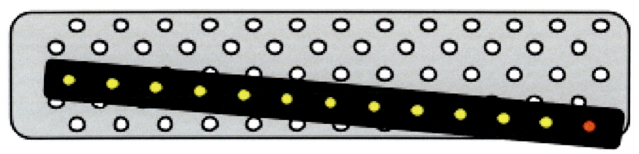

Fig. 6.1 Trial and Error Technique for Finding Non-Linear Alignments

Chapter 6

Though this practical solution works, we need a more mathematically sound approach. You can get help from that old standby, the Pythagorean Theorem. The way you use it is to make the point of connection for the gear one of the vertices of the hypotenuse of a right triangle. The other end of the hypotenuse is at the corner of the plate as shown in the figure below. We call this corner hole 0.

You can then use the Pythagorean Theorem to see if the hypotenuse is a whole number or close to a whole number. If the hypotenuse is close to a whole number, that means that you could fit a gear in that hole that would mesh with another one at hole 0 (directly or through idlers).

The next question is how close does the hypotenuse have to be to an integer to be usable. After a lot of trial and error, we came up with a viable zone of -1.0% to +2.5 %. The tolerance on the undersizing is lower than on oversizing as Vex gears do not operate well when forced into each other. On the other hand, when oversized they function adequately (though inaccurately) even with a good deal of play. Loose meshing creates vibration and noise and teeth sometimes slip.

Fig. 6.2 Tool for Finding Non-Linear Gear Alignments

Complex Gear Systems

This approach is best implemented using a spreadsheet. Table 6.1 below shows the lengths of the hypotenuses for various combinations of holes. The colored entries are the ones close to a whole number and can be used to fit gears. The green highlights show hypotenuses that are within 0.5% of a whole number. The yellow show hypotenuses that are within 1% away of a whole number while the red highlights show hypotenuses that are within 2.5% of a whole number.

For example, in our table, the best fit would be to fit a gear four holes from the top and three holes from the left, which is actually the only combination of holes that fit a gear perfectly at an angle to the main grid.

Table 6.1 Calculation of Deviation of the Hypotenuse from Whole Numbers

0	0.5	1	1.5	2	2.5	3
0.5	29.29%		20.94%		15.02%	
1		-41.42%		-11.80%		-5.41%
1.5	20.94%		-6.07%		2.82%	
2		-11.80%		5.72%		9.86%
2.5	15.02%		2.82%		11.61%	
3		-5.41%		9.86%		-6.07%
3.5	11.61%		4.80%		-7.53%	
4		-3.08%		-11.80%		0.00%
4.5	9.45%		5.13%		-2.96%	
5		-1.98%		-7.70%		2.82%
5.5	7.96%		4.99%		-0.69%	

A VexIQ Gear Handbook

0	0.5	1	1.5	2	2.5	3
6		-1.38%		-5.41%		4.17%
6.5	6.87%		4.70%		0.51%	
7		-1.02%		-4.00%		4.80%
7.5	6.04%		4.39%		1.18%	
8		-0.78%		-3.08%		5.07%
8.5	5.39%		4.10%		1.56%	
9		-0.62%		-2.44%		-5.41%
9.5	4.87%		3.82%		1.77%	
10		-0.50%		-1.98%		-4.40%
10.5	4.44%		3.58%		1.88%	
11		-0.41%		-1.64%		-3.65%

Chapter 7
Conclusions

It has been a lot of fun for us learning and researching gears. We have learned how to build more compact gearboxes, so that we can build gear towers more quickly.

We have also learned how to calculate the gear ratio and that if we have too big of a gear reduction (a lot of torque) the arm will go up and down too slowly. If we have too small a gear reduction (too little torque) we would not be able to lift the object.

We also learned that two stages of gears are almost enough for any gearing, unless you want something extreme, like a gear ratio of 125:1.

Finally, gears aren't just for fun, they are very helpful when building robots.

> We have learned how to make compact gearboxes with just the right combination of speed and torque.
>
> We found that two-stage gearing is adequate for almost all cases unless you need a gear reduction or gear ratio of 125. Such cases, requiring a huge increase in torque or speed are rare.

Appendix A
References & Sources

A. Selected Web Sources
1. The VexIQ Resource Portal – Unit on Gear Ratios
   http://www.vexrobotics.com/vexiq/education/mechanisms/gear-ratio
2. Gear Chains and Speed -
   http://www.robives.com/mechanisms/gears#.Uv0pGfldV8E
3. Relationship between Torque and Speed – California Science Fair Report
   http://www.usc.edu/CSSF/History/2002/Projects/J0209
4. Types of Gears
   http://www.gearsandstuff.com/types_of_gears.htm
5. A description of the Chinese South Pointing Chariot
   http://www.lhup.edu/~dsimanek/make-chinese/southpointingcarriage.htm
6. A Video on Kinematics explaining Involute Gears
   https://www.youtube.com/watch?v=Dh83mGUCiws
7. Ronson Gears' History of Gears
   http://www.ronsongears.com.au/a-brief-history-of-gears.php

Appendix A

B. Books

1. Maitra, Gitin M., *Handbook of Gear Design* (McGraw Hill 1994).
2. Jones, Franklin Day & Ryffel, Henry H., *Gear Design Simplified* (Industrial Press, 3rd Ed. 1961).

C. Picture Sources

1. Clarke, Arthur, *Worm Gear and Pinion*. Digital image. Wikimedia Commons. Wikimedia Foundation, 10 Jan. 2011 (Public domain image retrieved on 15 Mar. 2014).
http://upload.wikimedia.org/wikipedia/commons/7/75/Worm_Gear_and_Pinion.jpg

2. J.B. Lippincott Company, *Archimedes Screw*. Digital image. Wikimedia Commons. Wikimedia Foundation, 18 June 2007 (Public domain image retrieved on 15 Mar. 2014).
http://upload.wikimedia.org/wikipedia/commons/5/50/Chambers_1908_Archimedean_Screw.png

3. *Heosemys Spinosa*. Digital image. Wikimedia Commons. Wikimedia Foundation, 17 Oct 2012 (Public domain image retrieved on 20 Mar. 2014).
http://upload.wikimedia.org/wikipedia/commons/b/b7/Heosemys_spinosa_Hardwicke.jpg

4. Burrows, Malcolm and Sutton, Greg, University of Cambridge, UK. *Issus Coleoptratus' Gearing Mechanisms.* Digital image. Smithsonian Magazine. Smithsonian Foundation, 13 Sept. 2013. (Retrieved on 15 Mar. 2014). Used with permission of the authors.

Appendix B
Online Source Data

A. The Gear Reduction Spreadsheet

You can access the Gear Reduction Spreadsheet as a published web page at http://bit.ly/gear-book-data. It shows the gear reduction for each possible 1-stage, 2-stage and 3-stage gear permutation -- there are 819 possibilities! The spreadsheet also shows the Minimum Spanning Beam and Stack Height for each permutation. It also has the source data and analysis for Figure 5.3 (on page 40). You can also get to the source data spreadsheet by scanning the QR code above.

ABOUT THE TEAM (APRIL 2014)

(From left) Matthew, Angela, Rohit, Christopher and Megha

Matthew Cox is 13 years old, in seventh grade. He lives in McLean, Virginia and attends Longfellow Middle School in Falls Church, Virginia. He has been to the Virginia-DC FLL State Championship twice and is headed to the Science Olympiad Nationals soon. He can synthesize hydrochloric acid, scaring the rest of the team. He likes playing with chemistry kits and has watched all 5 seasons of *Get Smart*. He has won several elementary school spelling bees, is fluent in Pig Latin and Ib, and has visited 45 states in an RV. He even experienced a flash flood himself in Manitou Springs, Colorado in 2013. He wants to be a scientist when he grows up.

Angela Wei is sixth grader who attends the Potomac School in McLean, Virginia. She is 11 years old. She has lived in both America and China and is fluent in

About The Team

Chinese. This is her second year of competitive robotics but her first year with this team. She is interested in science, technology, art and literature. She has been studying music for 8 years, piano for 7, dance for 6, calligraphy for 5, theatre for 4, been a Girl Scout for 3, robotics for 2 and violin for 1. She has won awards in calligraphy, essay and piano competitions. Besides robotics, her other hobbies include coming up with stories, doodling, writing songs or pieces, finding science jokes and reading dystopian novels. Last year, she illustrated a children's book (When Disasters Come Your Way) on natural disaster preparedness.

Rohit Narayanan is a sixth grader who attends the Potomac School. He is 12. Born of Indian parents, he has been interested in math, science and how things work since an early age. In 2011, as a member of the **Pi in the Sky** team, he was one of the 20 World Finalists in the Google-LEGO-XPrize Moonbots competition. He has done four years of FLL competitions and has been to the VA-DC FLL Championship Tournament twice in a row. In 2013, Matthew, Angela and he won the third place Championship award (out of 530 teams), making them eligible to represent Virginia in select national and international competitions. Apart from technology, science and robotics, his interests include writing plays (thrillers, especially), theatre, opera, traveling at home and abroad, writing a travel blog with his sister Megha (at http://traipestry.wordpress.com/) and swimming. This is his second book with Angela and Matthew.

Christopher Kang is a fifth grader at the Potomac School. His middle name is Korean and so is he, though he was born in Fairfax, Virginia. This is his second year of robotics but his first with VexIQ. He has been waiting to compete in robotics for a long time. He has been playing soccer since he was two. Other hobbies include piano, clarinet, math, reading, writing and building. He also enjoys CAD design. He is very excited about reaching the VexIQ World Championship.

Megha Narayanan is a third grader at the Potomac School. Her father has been coaching robotics teams for many years and she has been looking wistfully from the sidelines. She is really excited about her VexIQ debut. She plays the piano and goes swimming three times a week. Her favorite hobbies are inventing things, making up jokes and reading. She created the Professor Gearhardt character.

McLean Robotics Institute

info@mcleanrobotics.org

Fostering competitive excellence since 2010

2011 Google Moonbots – Final 20 Worldwide
2011 FLL Regionals (Fairfax) – Div. 1 Champions
2011 FLL Regionals (Fairfax) – First Place, Robot Design
2012 FLL Regionals (DC) – Div. 1 Runners Up
2013 FLL Regionals (DC) – Div. 2 Champions
2013 FLL Regionals (DC) – First Place, Robot Performance
2013 FLL Regionals (DC) – First Place, Robot Design
2013 FLL VA-DC Championship Tournament – Third Place, Div. 2 Championship
2014 VexIQ VA State Championship – Overall Championship (Excellence Award)
2014 VexIQ VA State Championship – First Place, Robot Skills
2014 VexIQ VA State Championship – First Place, Teamwork Skills

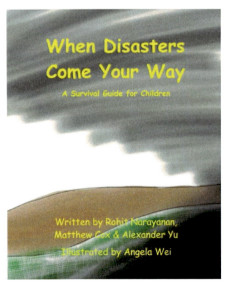

When Disasters Come Your Way: A Survival Guide For Children
By Rohit Narayanan, Matthew Cox & Alexander Yu.
Illustrated by Angela Wei

ISBN-13: 978-1-493-78071-6

An illustrated children's book on preparing for common natural disasters. Now available on Amazon as well at your local bookstore. See http://bit.ly/disasters-book for more details.

Made in the USA
San Bernardino, CA
25 October 2015